W0115275

OTHER BOOKS

All Night Salt Lick, with Laura Chester (1971)
Solid Object (1972)
In XX Arrondissements (1974)
Mixed Doubles, with Artie Gold (1975)
Subject to Fits (1980)
Elegies (1985)
Rocks and Deals (1987)
Pockets of Wheat (1996)
Admiral Fever (1997)
Rejection, with Michael Gizzi (1997)
After the Fact (1998)
Space Jam by Billy Higgins (1999)
Able Baker Charlie (forthcoming)
Drive, It Said (forthcoming)

#42083945

PS
3575
0789
C47
1999x

99-10-14 dlb

CERULEAN EMBANKMENTS

*

GEOFFREY YOUNG

DRAWINGS BY CARROLL DUNHAM

LIVING BATCH
1999

OHIO UNIVERSITY
LIBRARY

WITHDRAWN

Cover painting, "Cerulean Embankments," 1998-1999
and section drawings, 1998, by Carroll Dunham

Thanks to the editors of *Lingo* & *Mirage #4 (Periodical) #71* for
publishing versions of some of these poems. And acknowledg-
ments to the slack-tuned lyre of Bruce Andrews, the proofing of
whose galleys for *Tizzy Boost* launched the writing of this book; to
Robin Young & to Mike & Malik Solomon for early audio recep-
tion in Sarasota; to the breakdown of essential protein chez LH &
to the "mille phantasmes impondérables" chez TB; to the screened
porch, green table & summer dawns at 5 Fern Lane, Woods Hole;
to the protean voicings in the original score of the weekly Gaboon
Viper show; to the great variable foot in Sharon's long march; and
especially to Michael Gizzi, whose generous sessions of close read-
ing helped shape these poems.

This book is dedicated to the memory of Don Cherry's emblemat-
ic volleys, intervallic leaps, duet harmonies & constructed bop edi-
fices on pocket trumpet and cornet, as well as to the "compleat
angling" of fisherkings Lee Morgan, Booker Little, Miles Davis &
Clifford Brown.

Typeset & design by author & Chad Odefey
Copyright © 1999 by Geoffrey Young
ISBN 0-945953-10-0

A DRIVE HE SAID BOOK
Living Batch Press
518 Hermosa SE
Albuquerque, NM 87108

TABLE OF CONTENTS

CERULEAN
EMBANKMENTS

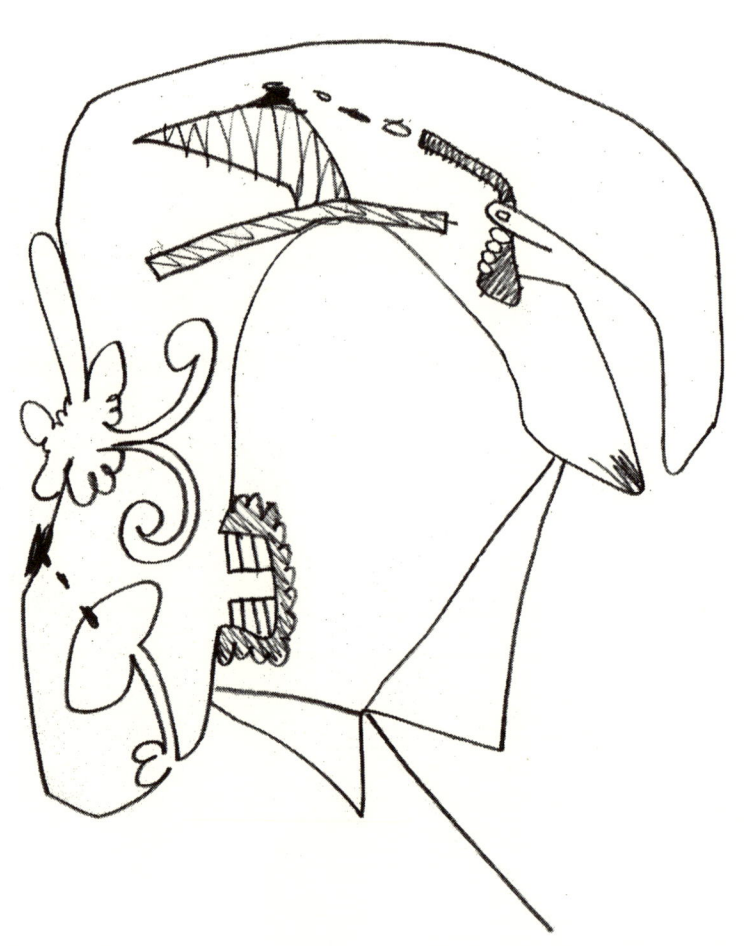

THE

KIND

OF

MONEY

THAT

MONEY

CAN'T

BUY

1 Set writing dial on "emotional life"
summon jitter blood from effervescent lust
      —comeshot throated lyricism,
      meet white knuckle verse—
          to suck the lip of exquisite obliquity
"Carnal" works
pour "pleasure" tipsy, make *seizure* frisky,
      while Peggy Sue mounts Pegasus
to fuel our midnight maelstrom lamps
      with ear gold otherness
brought up from dotted folds,
      sleek-furred fantasy
          landing like a footnote on a minefield
No class assumption cuts a swath
      across these prairie vowels
      with herbicide,
on platform heels in hospital gown I.V. drip farewell
      to night's larynx venting an inhuman valve,
because our fleeced suckers flip and we're doubly wet
and proud to rehearse "Hello, Dolly"
      before phenomenologists
It may be hard to take this seed-state
      and turtle wax it with suitable lungpower
      on goon-tunes

that, once polished,
writhe in fatal traction
but all soul survivors
feed tokens to the spirit trolley
electing to ride out melodies
in thought balloons
A body's a mind made up
of playgrounds & citrus rinds:
prepare the coming coolness of optical koans
Now with chopsticks I detach "rust" from trust
and sip the tea
(pith owns pants that adultery riddles off)
cathecting Broadway neon in gold lace arcs
that peel steamer trunk decals from the lexicon
Let me retouch
your pronoun's exuberance
with charitable fingers.

2 I know questionmarks rake your speckled flanks

but on the retort mat your breathy reports scat

               Lee Konitz fragrance

                    to helm keening cool

aiming Josquin flashlights

     through church tone caverns

     now that habit swelter invasion is here

          teaching us fly-by-night staircase wit

     as song standards home in on the home range

     circulating pricier highs for daintier thighs

               (now that face to face)

                    in rooms enlarged by trap doors

meaning blinks from "I to we" like sibylline decals

          peeling from a pup tent

(inside of which our pep was spent)

          "I'm ready to accept cohesion lessons"

    To bend spoons

      on the bottom of fruit jars,

          use the slab-quickened 3-D blues

to eviscerate "without" by staying wryly "within"–

             quel skin!

not to excuse laminators their glue-happy pressing blurs

but to tie one on

      like cornet gods just saying no to milk

           in half-valved neck-puffed hours

           on jobs at chatter inns,

     a regular Dutch still-life

     of peeled lemon untouchables singing

           "Look at us!"

           "Our haves have not!"

     A day to hose off on.

3 Dabbling underwhelms the hard of art

though hot spells grab at nouveau intersections

    where verbs dispute all rhyme-free themes

    and the tube shrinks from sense

    as music grips a file & rasps: speed links volatilize

        (ornamental gestapos spank practice)

Names like Rabelais milk investment schemes,

            matter rejects Ur-claims,

                and adopters abort parents

            since charity flies eye-to-eye high

There's only so much mush to eat (and a hand

    beckoning from the robe of a swaying body)

        for my "I" to beam at pulpit concessions,

    blazers on angels that produce by-laws for in-laws

    in line for halos. . . This year's arpeggio ovum skinny?

            Any conception so ingested

            gargles things mercurial,

bottles of empty standards sit dewy-eyed in stitches

            linking all dearly beloved habits,

    but will the law defend our sometime exact lack of *sense?*

Guilt pinions curse then clam up

        to free others so afflicted, & by others

                is meant our bite-sized

    media lax consumptives,

    the oracular deans of this year's dark procedural grafts.

4 Peddle "panic maintenance" to hysterics for pin money?
   Paper your vaccine barque with yeast presentiments?
   Maybe we're too amalgamated to bake themes in unison
                  with seconds of sloppy say-so,
                      our love's bodybags filling with how-not-to
   lavish idle fire on consortium sports or court revels. . .
            Exceptional how daft is cult delectation
            torso by Pompous
      affectation carved from Carrara slab stringencies
            but when I get a load
            of your jolt suds peering over mole summits
I know why our chief executive artificer sits staring pert,
            a truss curbing any who would help crease limits
            on these stubbed cerulean embankments
      If I call for you Friday will you pick up?
      Mental purrs surface through riot quiet,
            set keel to breakfast on sidereal floors
                  Inside cognition's nerve terminal
      pride is ghosted,
            a length of won orgone on some azimuth of its own
            flips this week's lip allowance
                  if like quiescence what spooks your
                  portable anomalies is haiku wholeness
   Battling radio talk

spawns game board station logic
painting the supremacy of advertising stars
on a windless sky of no absolution
Every soul submits to twenty questions
Appeal is giddy as I step forward
to send Atlas an anklet,
lease Até an organ donor outlet

But I'd like to order something cool
from Merrill's good gray prison house of bird tropes,
from his Dalmatian butte overlooking hurt
where to sing out is to flow home
at last
to wind up patternless prairie
But thigh, to thine own
stairmaster be true
and it, the horizon, will bend to you.

5   Your fine thing may be chop-topped
with twice-pipes & chrome dump hanky pants
but I'll never again kiss
                  your mainframe's sapient amps
nor front croutons for your soup-city's zen depot
"Single female age immortal seeks single man
                  pretending to well-read scheme
            for adventurous interspecies travel. . ."
Might I be her crazy caller with flap epaulets?
      My antsy constructions rise
from no esteem to low
to maniac glow via serotonin's loquacious drip
with immunized page turner genie-free guile effacement,
      my society's any pair of imp ears,
                  paint eyes, or lip balm you got
      kneading the arch
            of a bare foot

      So may we be
      twitch-focus earthquake makers
rocketing at cannister speed toward galaxy film,
            at least until our ancient fear
            gears down
                  that last bionic offramp for cybergas

—a charade made motionless

        by legs glued to S.O.S. dreams

Thus my stance is convict spoon

        rhythm protest,

"old glory" blood-cashiered

        to write off fratricidal drugs

        buried in home clues

    of "honor thy offspring" consciousness

Here, cash peerage,

        crack this bluffer's tottering safe.

6 Dejection droops over weather-charred Elba,
carving desktop infarcts in exile cubicles
cliff face up, thinking "who'll die first?"
We who redraw pictographs as assorted hangman dreams?
My day job churns
bare legs on the dead run
rallying sun ducats on ice water to reblock love's beret
Chihuahua!
Give me sounds so semantic they yodel holy modal schist!
Give me what I miss, sis
your soul's privacy
playing the minx in moonlight
when with bright eye-lulling fancy you darken
and we tryst in barefoot loco weather
Only you
can bruise the silence with yellowjackets
as your humidity
in felicity's pool cools
a buttery stack of tannish stem analogies
punching my folk apex
with bestseller-to-be memory
But let these back pages dine on proverb deposits
in high-interest plots:
no *jeu d'eau* will breed birthday cataracts
over our dead body.

7 Faith wires residential minds
       by transmitting
   nipples as substance
                making ideas disfigure
       appetite
   (the way wind-stiffened flags gouge signatures)
Because oil's a fact
           that hegemony refines
that's right, I'm offering a case price
           on Rilkean has-beens
       to catch knots as arduous years
                I'm part head, part
           a hat's viscosity
parodial view of the meek's situation
as something justice fails to do
           but I won't generate misnomers
           to caption a peach
If our electronic peers spray the electorate
       with figure/field repellent
(pushing theoretical optics
       to preempt invisible events)
                it must be ordained that we sleep
       the sleep of subjects.

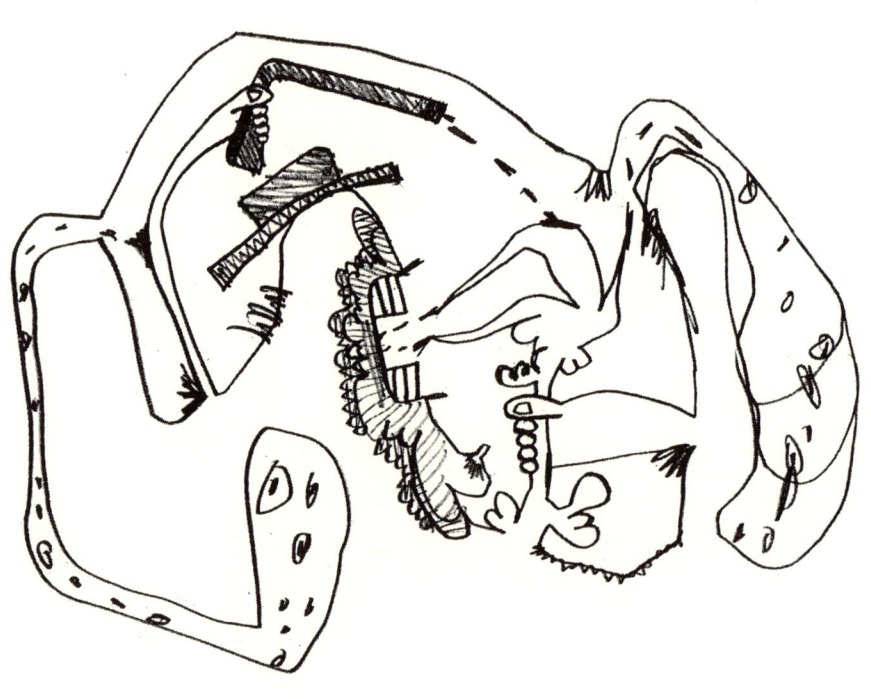

ZULU

BEADWORK

1 Cork walls bobbing on sound. . .
Beds desanctify poetry,
stir the incessant "Honey, is that you?" pot
of cells listening for elves through time's
spotless window
"Uncle Hate's is a fantastic mai-tai place," he said
(comet conjecture is to seed catalog agronomy
as cadences delivering pineapple curtsies are to. . .)
but can you
halfway-house the Attila in you
to pitch halcyon dice, seize golf cart batteries
to break self-destruct lineage?
"Coyly," *froideur* shrinks body
a kind of twist-off head decision, end of
chapterish, a weak sport in empty shorts
So let's get right
adhesive, be
trapeze grasped, know "gorgeous"
in pearl-toed pumps, this sticky
autobiography of no one spelunks
"Is the grocer a grower?"
Something you tell me spells
"strangers who belly past bright windows"
the ones who make wreathes of coal to reheat spectators

or pitch monthly Imams with

E.T.'s junk phrases

etched on wood-panelling's inorganic sheen

Rock's right hand

plucks a touch-me-not string & hears

cash rend the qualities it banners

but is not our vehicle

tenor enough

to exit

the tune wreckage,

melt rings off glass hands, break brains together

on our knees?

2 Dressed down from angel to woman
for the scribbler's Bestiary Ball, you're insolent
        with respect to first-aid partnering ink
choosing Cole Porter's Heidegger over Buster Keaton's Budweiser
        Like very impasto, your oracle: "Why not jack up
                God's slang assailants
        with a lassitude tax, unfurl His sugartown idioms
                against frenzied mercantile boasts?"
        Left holding the broom I'm sweeping
                our personal stats into time's canvas totebag,
                a Ball with seams scuffed along ravelling dreams
Whose career-year will dare folly, or nettle impact?
        We may be short the air fare,
        our market share reflexes lamenting
                "Homeric Lovers Whiff on Breaking Stuff,
                Join Pile of Heroes Stacked in Lost Cause Column,"
but when late night history summons your lemonhead gray eyes
                to the naked-on-a-green-sofa derby
        Circean chemistry puts improv knock-out drops
                in the evening's "Falafel in C Ya" horn solo
that we might overcome fear's smoothie butter pats here
        A striation sunrise's golden win-win granulation
                is color enough to race in mind
                reaped bounty   Even khaki, even tacky.

3 Objects harness desire
as prescient collectors acquire,
        but bend to pick up a glass slipper
from headlight streaked midnight macadam
or hike red hat
through property shackled with fences
            and you will sense impermissible goals
                routing the town:
my "yield machine" steals sweat equity
            from the poem's projective field of dreams

                But whose accent
        will legislation repatriate?
A poor face is not a spoor face, Elvira
        Then brightness enters switchback
        on blinding bel canto overbite tracks
            fretting in half steps
to spackle my latex gloves with downy heiress dreams
        "Venus Mouth Sings Male Lead"
            the headlines crow,
                subtitled,
"No incidental moxie powers this penitential vow."

4 Sound the countdown to no-count sound
             cupped in parenthetical armor. . .
   (Dash hard to planet's edge)   (Speak igloo)
       I'm table adjacent, you're cortical lips,
    viscosity depths in sweeps of "here, no evil"
                 with fate olived-out on toothpick waivers
by dint of po-biz italic eyes
         while Childe Harold swims out
     beyond non-font cock games that screen bangers seek. . .
   Maybe I'm a mountain pass
         maybe a body frozen in ravine ice,
     a refugee stalling to eat
     himself, acknowledging the nightmare
             flight of your shoulder-strap *ciaos*
Literary forgiveness starts here
         "on the premises,"
drawing critical blood from brocade inscriptions
             in starry fuse fame
       because letters repeat & repetition communizes
         & a golden bump scene guarantees a creased Diana

       If we twist phrases (lather gesso)
         into polemical victims (heroic alchemy)
     a liquid crystal surface deals *caritas* in bite sizes

We breathe lawn esteem,

we mallet filigree trolls

dressed in semantic quick-release sweats

lined up for eager ovum contracts

that bind us

like stewed physics

to this *terre inconnue*

pot-au-feu.

5   You glimpse Daddy Raw glomming art smarts
from sleaze please bulimic midriff crunchers
biting the air of cheesy numbers with Swiss Miss persistence,
only to realize The Movement to Respect Compulsion rejects
"respect compulsion," calling on aesthete wave insight to heave
trance place-names at revisionist terms of Beat's badboy bed nativities
          while Republicans channel bit-part bank tellers
          to deposit furtive disrobers of nightly demurral
          (only the Muse *steals* lingual ice
                    from topers of the Round Table),
& Dems pose as rapper stowaways hiding in media
(it's so *genug* brittle to brag of fellaheen flappings
                    or flick your inner bystander),
          their propane laptops blunting Byzantine soup
Because nothing means merger to a skull merchant
          if in gangsta your ism risks losing its blues
          to farce, mob acknowledged cortege reigning over turf gold

          My baby's text meadow is an election veil
whose jape honey legit hooks my hospitality
          with what she slings to taste
                    (the lone-gene star of all rinse lustre),
          erupts in pickled jams
          of spit crack matinees.

6 Skim-lip backwards by milk light

where "no limit" nuzzles white friction

But is it better to disrobe collectives in nearby nests

than pragmatize spokes to fellate hubs?

(This is shown as errorless spelling. . .)

I am forced to corner purity

with a fire-poker

as even these drafts do push-ups on stinging embers

because

items marshalling remote halters pledge

tan-lines on omphalos piercings

All scans stream, so what if they're asphalt prissy?

Behind clichés

the fig leaves opt for naked beacons of river equivalence

fooling no one

as to rayon's glissando privates

So let's walk past colorfast

nativity shrouds

dampening power profusions

with shivering reproach

Agree?   Or greed?

A pen am I, a motive swan silvertone

durable dream paradise streetcorner

with speak no circuitry diamond

coital lashes

I know you know how much

your hip cavity pulses

re sheerest camerawork

tunneling all craves

(You're as perdurable as moping is mawkish)

A strategem

of phrasable auras pays:

test best

to stir best

slogans success.

8 Reprisals steal by night
          luring each electrode to the jelly
              "Genital One, meet Anger One"
because all good galley slaves on the Love Boat know
     épater's another word for bug
          and bug's another for bounce
I wonder, is there a cosmic physiologist in the house
recruiting our neurons,
          ready to snap us
              like "nut 'n honey" lab mice?

Bad things like mirage water grow in us

              What is meat damage
              but withheld food,
the fuel all bodies are hostage to. . .
          while "lapidary" sing the beverage bubbles,
          tepid Rome petting Pontius Pilate's boner
              magnified by sorry hindsight

              But I should care
if your diction contradicts chimney sparks?
If star-gazing goose-down specialists
          cross doxy swords

at a Cornball Banquet?

Friends' eyes

spin in bite-sized blurts

investing their flesh with scheme inserts,

a ragtag band of flatpickers plugging in

to sightread leadsheets,

singing **Drink Kava, Now Forage,**

Ma Fois,

For Propitious Poetry,

& **Extol.**

SOFT

LANDING

ON

HARD

DRIVE

1 As pots to a chef,

sombrero to shaved head shine:

I like to say goodbye to people before I say I'm fine,

shift touch to finger tips

that caresses may rekindle a de-spleened

eagerness for syrup verbatim

boasting live mammal gratification

joined at the hip eyeball steam

lifting lithe diva deposits

to the heights of Mount Hereabouts

But when fun culture sickens

as knowledge goes

nowhere, observation

posts relay Hollywood levels of gonad toxin

from pink-eyed sycophants under house arrest

and ordnance grubs

in weapons clubs

sell gat adepts

nicknames in limerick charms

Who ain't got a vial of *Dial Nervous*

to own his own lucubrations,

gauze electrons in borealis troughs

of flat sun shards called poésie
This year
of impairment proof
bone crucifix madonnas
we're calling all
"step up to the plate" saviors
to raise the roof.

2 Ladling "PURE BEAUTY" soup from Golden Mean Bowl
            I exit with a bucket and a jade lozenge on a chain
            softened to frame
      abruptly lyrical footprints
            in mud as told by Mao's legions in T-shirts
Reliquary-sized children in moghul bee looms
      suckle the fruit of advance wannabes
      —hen in coop lays one gallant Fabergé—
      but here's my ice, my alloy, my mole fabricator:
            "Attention is rent by embraceable alerts"
            We are bodies
            only by being trapped in borrowed letters
      filing the affronts of art's howitzer critics
      as they set out for Ciudad Twombly on orders
            to disfigure the self-amuser,
to stain   scratch   abrade   bleed   pistol whip   or mime
      his blackboard turbulence with fax-heat repetition,
      chasing his goop baby paces
            through ideologic misprision
      to leave their incommensurable palaver
            like burnt offerings
            near his lumps of kicked up
            whale barnacle motleys
               Oh the handsome things
               another naysayer says.

3   They pull the ripcord
            on your marquee value
            your fatburger runts perish

        (Nomenclature meets Tough Love
            as Obliquity
            fights Encasement.)

4   Junk ads rake our tribal days
　　with a grifter's gift for slipped skin bunkum
while on ghetto shores insider muscle
　　　　　　　pothers incestuously,
a fierce narrative onramp sparrowed by snowflakes
　　　　　　　wings in on its inward "spiral"
trip-switched by Psyche to a three-piece suit
　　　　burning privileged "I am" price leavens
　　　　　　　with rivetting eloquence on Tanqueray rocks. . .
But is this brainpan a sink in a souk gulp disorder
　　　　　　or a lay-up rolling around the rim in spin?
　　　"Gun the pun" for manumission fruition,
　　　　　gentleness the knowing not
　　　　　　thinks asking "So What" of "Too Much Sake"
　　　as mike-master Miles & finger poppin' Horace
light out for the original territory in color-free
　　　　　　　bone-handled spoon barques,
　　a present sense adventure in mood solo terrain
　　　　　　unlocking Fahrenheit pinecone sap
　　　　　in a sweat drenched safety net feature
　　　　no army with stripes would ever fold into
　　　　　　　its contrition bills
　　Here's the warrant: eat micro-world
　　　　in the hard beat's
　　　　　selfless hour of let's go.

43

5 Else how but on?
Elsie wows in green
Says who Hell's not wooed?
Eyes the onset owns

Else how but in green
Elsie wows as one
How else but hooked
Touching her bare book?

On & who
On & green
Out & else
Hook & dream

6 Negotiate rapids laid out like marzipan

or pull the cord on fidelity's chagrin drapes

to reveal the plateglass of a no-exit landscape

as our face to face grindings play up the funny plings

and shock mediates a written soul's clandestining

like hosts in ties more ginned than smooched

filing future names in havoc hissing bounce

Love's brain melts microfiche, a reckon refreshened

rectally retinal reminder

to summon fuel worship,

but will this fire law douse cramps or blister art

in realms of revealed Picture?

Bambi, Bluto, Bucephalus or Babar?

While cyberphunk beats drop

students of ethics shoplift

and atonement shrinks a sham

why not hoist these chewed up animal bribes

(we offer dowry infants in lieu of serum slices),

to read blade muscle as winsome gradient

My arterial craqueleur is garlic tongued

confidantism in luxury boxes,

a cold face surfing as if

death & beauty were a mirror's mercury

Fingers greet swigs

I promise raiment, a truckbomb of rapid transit cuticles.

7 No sound is dissonant which tells of regular life
      but when you cry with pleasure in Cape Code
            your giddiup precious
            waltz inflection tears
      body forth sweet face-quark serenades
                  like lapdance sonnet needs
            dogging right up to the moment's
      priceless edge to declare:
                  "feelings themselves have feelings"
            swallowing our eye widths
                  in the dark of blue
            hydrangea streets

So now we sport Natty Bumppo blinders
      and hobnob with racehorse
                  pollinating grail harems
      quick to win ark pups on dime shrugs
            until a voice says
            Honey, hear that stagnant buzz?
Must be a tinhorn president piloting vagrant fuzz
      as your lean leg yoga rivers me alpha omegawise
            sending my image chamois out
                  on dampened shroud duty
            in the taste massage

& boomtown gone bust
toggle switch nearness of you

which leads me to point and shoot this camera
at genius
storefront windows displaying
sting-ray shirts
against pale museum pride,
because in this scheme our star-crossed rakes
heap leaves as crowns
to answer answerless time!
preferring "the same, only different" finesse linen
to pre-emptive oxygen strikes
until all becomes rhapsodized alertness. . .

But as bud follows mud
and October September
so like spoon shine I licked your superpower
before the great inexplicable brushoff
and now time is nothing
but thinking I see your car
when I don't
Cold noodles
under roof tiles hinged to sun.

PARTITA

FOR

GLOCK

&

HOUSING

STARTS

1 Shine, little tremor,

on any me revealed

within any you as field,

this co-delicacy beneath foam

brushing belly aloft in drowsy ease

curbs tongue nerve, leaves

mind adrift in come-from-behind lineaments

that jimmy poetry's long limbs

to new levels of id sire,

implements designed by Cry-no-more Samplings

glueing chic cut-out atoll shapes onto tall foreheads

as this passion for Cupid puckering

fades to beehive sleep

(Victoriana makes Bedouins chill)

Blame Venus if you dare, but a love story

in ruins will peel a lifesaver from your tongue,

its horizontal pilot on morphomatic,

a script-coup issued by stuntmen

shrinking all "we who seek opportunity glow" types

to calendar stills etherized upon an angel

That's why I temper my waltz residues

with no-swoon congenial flops

even if figurine cities switch brooms to mops,

my cell exits are buffed with

their own cryptic inscriptions:
"words unsaid need no revision"
craving unbruised tubes of poster bodies. . .

Now a collaged eye is
the organ it thinks in,
laugh track lines
sobbing hot holistic groans,
my new shoe
in your super glue.

2   Falsehood: let me return your "alrighty then" welts
        to arms patrolling stealth-eared rabbit sentences
forthwith cemented to paper
            by nerd saints choosing predestination
                that no more boy toys
      starlets grope for
      will lie face down in pools of myth
                to exclaim their incontrovertibility. . .
      (But why describe these shotgun-in-mouth detours?)
          Newk's looping Satie discretions
              amass boilage
          as Sir Philip Sydney's "I I oh I" courts inanity
My old flame's a binary germ looming to be torched
        even as day to day coolants chill her fabled Siberia
           but when sudden species take hold
      we revise connections
            we're part sweet, part strapped to the stacks
            we are vacancy itself, a pawn to ethical shouts
      running like goons on shining angst
                If affection points past audience
                fingers touch eruptions
making me print "the terse, vigorous scars of prose"
        far too esemplastically for right angles to be wrong
        My schoolgirl diet's an inferno, I must
                    lick this first-class stamp. . .

Then her secret polevault epistles

are delivered like bonus bamboo

      harpstrings, leaving me here to creak,

a windswept valley

      designing the new nonchalance

When it comes to lust intake

we dream too much "hot to"

not enough "snake total swivel labels"

      in this our digit practice

          for summer's bueno vibration seraphics

when pointing to the moon raises smoke fluency

      to new levels of tinctured sensate leap signs

Can you tally sums in sequence, as in whence becoming comes?

Being I know to be all

hum baby seepage purring within

      little screens

Dear Ultimate Concern,

Please favor the crown of the babe

      pushing through birth canal,

      & because pertinent,

          give theism a sitz bath

to breach this brazen two-ply gaze.

3 Sharp as fake love

> mouthing brainwash twinge

> my purgeless modem

>> disconnects Art's

>> "I questionmark death" repetitions

>> to steady the pall

while enslaved flesh obeys liquid protein

> scoring fold lines

>> in our search halves

> Time steals all runes,

spreads *bobkes* on bad dream crackers,

filling green hats with randy opera swag

The protective coloring of a glittery wingspan

> will cloak a nation's trigger finger,

> a sucker wind

>> pitch a fakir plot

> to the tune of

>> "gags of all stripes ironize"

Hold no brief with, take no stock in, pen no ditty:

> the jingo-lingo for "pax americana"

>> is "business plan"

> as cell phone grenades

>> launch talk stampedes

while invisible suits

looking down

through plate-glass at the tinted sheen

of a silver river

smoke long ones

called NOW
OR
NEVER

Alert!

all ye mentholated nobodaddies.

4   Planted like a weed in Gethsemene's bed

I'm ready for her

star-quilt shutterbug raid

on the polymer

of sheering subjectivities withheld,

when I hear

"snap" as unstrung, "pluck" as unsung

O veritable nose

above silent tongue

is it

from zircon's loot bag

nest whores dish beet soup to unloved pukes?

When trust melts down

to pinpoint pupil glare,

distance blinds & words wither,

but let's not gainsay, misgive, or hamstring

the past as humanizer, *nein*

Strut your pronoun definer, survivalist

while I rub ointment

on the new ennui.

5 "Git holt" of your crashtest
                    infantile feelings, Little Dorrit,
or watch critics paint curlicues on your curfew symmetry
                    Suicide damns the almanacs
        but our rump extravaganzas mulch sampan stools
occupying a laugh script so creased with azure ingots
                    it's been mislabelled
              "cupidity's albino task"
                    taking serial monogamy to the absolute moot
        before switching digressions to
                    procrastination briefs, lethal
        lamentations of what's pigment, what's pie-eyed

                    Meadow sex in unmenaced aura
              sprinkles camera dust on razor eyes,
        art's strong-arm lies
                    waiting to decree that
        majorities chew the zweiback of limited phobias,
              their gene pools drained to the tiles:
                           a blink is as god as a plod
        in advancing skin cream diplomacy
              putting make-up on trade slaves
              before blistering bell curves to hasten
              a moribund penchant

58

A power-mad political amorality rules puritanism
     complete with "I'm sure" "I'm not sure"
      and
       "You say that now"

          But then
          I'm as cynical
          as you are dotage proof.

6 In a face-off of serial inebriates
Strife's ampersands connect digits to data banks
though we hang till ripe, flush with flesh-built differentia
eager to anoint new teeth with wee sentiments
or point grid beaks toward other suns

Because your complex demeanor's red hot and bothered
I'll be your ace townie surfer locked in mortal coil
shrilling my fluid registers, decreeing *rien*
but a yin privacy beats this quaint yang
pout of a patsy film, the one called **Gone Tomorrow**

But let us prime time's systematic thaw,
this "I'm not missing history" homily
in stained-glass sound,
as talkshow-lite-flavored immaculates
of hair streak opera tabs sing of surface despondency

Whose icon amnesia this is I think I know
shall pass through metal detectors
of annoyance to sanctum shopper appetite
as mall-scanned in pleated blazers
we extend a virtual hand

into the living bargain stacks
where Proust's involuntary French cock
nuzzles Ti Jean's memory babe eagle,
K's desk prompter the At-man,
channelling ionized Mary.

7 To be imprecise as voodoo
writing instruction manuals for unborn skills
(hangnail venom staved off thus
                    by graveyard distrust)
ignites the dredged particulars
                    of privacy's caramel confessions,
these transcribed lumps of planetary potato moon chat
                    bidding the text adieu
                    (that's what we do)

                    not speed-dialling belief to vacate inspection
                    or we'll bury Caesar the whole nine
                              Scribbling *this*
                              is pure sport
a flick of the wrist varnishes the spar, hoists the mast,
            takes on peat and billfolds the wind
                    since all this "dispensation"
            is just duty-free in between *mind* (the one
            that erogenizes the casket market, frisks
                    local understudies, closes up ten points
                    in arch relations to earn
the palmed-off anew), death beckoning a château
A scandal a day plaits
            an epicene beard, our prior

convictions exit

a tense so draped in past tinsel

as to reflect a narrative

that all our fathers juiced to, demon mist rising

before acid paragraphs

stamped the future "as advertized"

in consumer carts, a pile of jettisoned logos

whose job is to collect interest

Still,

incense ash

am I.

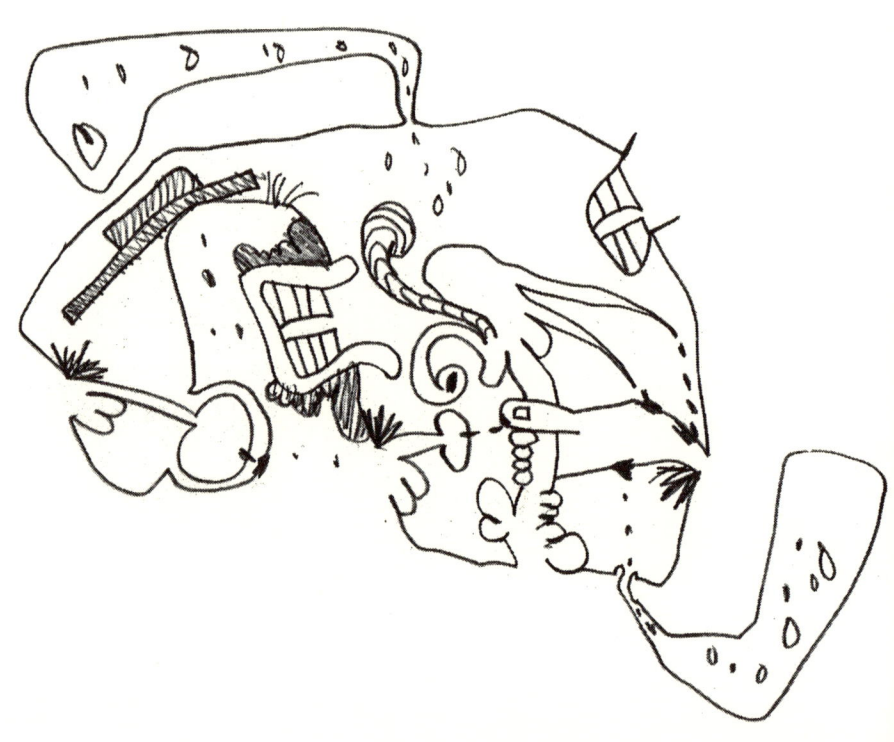

ABUSING

THE

DRUG

PRESCRIBED

TO

HELP

1 Scrape clown face off self mask,
graph master pilot carnivalesque wing dings,
                those boob-a-lot
loop-de-loop transit cultures
                chipping mileage steroids
                warped as the weave of a Trisket
        or the chemical basting of distressed organs in crisis
                if only to hail
        thumbnail sketches of air traffic control maps
Even these putter decibels
                simulate halter-topped History
when ageing forgiveness is urged to melt species ice:
        a smooth hand
                can free a unit's information

        We're plaid, not mad, glad
                to be unhappy
(thrones of stone
        thrown down
                public affairs stairs
                        fill Sun Ra's date palms
                                with Arkestra cuckoos)
        stalking zen's love-handle lemons
        as sentience dismounts

from thickened peeves
and sheds ripe cloth to reveal a scribe's bent back
about to harness gender chips
that switch breasts
to verbs

If every breath you take pollinates
as every cab you miss dissipates,
poetry's big screen muse of the moment
still mines the mimetic flash,
a plate full
of Isaac Babel beer and opiates
by Shackle, Tilt, and Pierce
Give your marbles
a spot
of tea, give
your roam the tone test
a new home.

2   Her big idea drink ship
pours thrill seekers in thrall to trash books
shots of labyrinthine ashrams
                    decked out in hair-pin firings

Because sound propels, & conciliatory air wings it, & a cig-in-hand
hyper sniper cover-up artist trains lipids to roll greenbacks
in "never enough" grateful sugar. . .

Maybe that's why they hired her trough trimming vex?
For outsider value? Spleen's judgments yodel elitist fandangos that melt
precarious jokes as Salomé directs many a lopped-off snuff tape

starring Creamy Material Run Out of Self-Expression, & now this
minutiae stuns art beacons pointing from eager storefronts
     at mispelled *potatoes* (the delicacy of the month flave)
& nary an "I shouldn't stare" horizontal brainstorm planting can stop it

          Light—a monumental abbreviation—adds
                    unknown red special lag anthems
                    to slur-by-night pebbles tossed at Windowgates,
          as Blue Meanie phobics in riot gear
                    stir up sperm dust etiquette
                              radar-screened to amputate
                    a nomad's contrary legs
          while we trawlers bottom out on light blues.

3   You don't have to be a moron to see
                    today's upshot wit going hatless
             via "darkling plain" jokes
             concerning skimmed-over loyalty negligence
                         or to feel in the breeze
             the inconstant heart
             is genetically fomented, each dart greased
                    by Cupid reporting an ache
             as commitments forage in over-grazed musings
                    that lead the blood on
                         to next wave garnerings

Knobs of start up instrumentation capture the brisk smog beats
        of "It's All Over Now, Baby Cakes"
                    as when
             Madame X slams a phone
                    and I fall on electrostatic knees
cropping debutante fricatives in moonlit section marks
        "I'm dry, I'm semi-dry, I'm soaking
             my ducktail in false memory
             to streamline fortune"
        Even bumped down to fuchsia therapy
                    we're not well
                    beneath onyx pushpins

The Sermon on the Mount

(the onset of Scruples in the Chest)

                  peels a pulp genotype

      as a cracked lip thinks

      I love the rarity of a familiar "it."

4   Despite Lucy Lipps' cheap shots
        I dream straight ignition ham,
see backyard apples
        leveling a juggler's gaze
watch health cops bracket
        the same soma
hermaphrodites crave
        when they shrink
their graphic waistline tailoring
        into body-zippered
soundtracks, whistling
        millennially. The truth is
I never left Argentina, either, Olga,
        so beware, oh my friends
of the melting pot word,
        of an egomaniac
with an inferiority complex.

5 Invent hindsight by screenlight
　　　flag flap over flag
　　but come hell or high Jesus
　　ain't Golgotha gregarious

over-looking a town without self-pity
this respite in "Father, forgive yourself" hangtime:
any sonnet scribbler's an empty room
　　of spade flush conceptual ink

a quality deeper than yam level gestation
　　　swears, "Your love allows me to quest"
even as miscarried words grind
　　　　　　their marsh breath

When "alls I know" quails
　　　onan station supplicants
　　　measure bed phlegmatics with calipers
for nerve-cured guest kudos

　　　hurling buffet wedding rice
at Portia's pink table accumulations
—a life refined by selection and arrangement—
　　　suddenly whiting-out

OK so maybe your genius *is* figburst sweet
but hard of heart's no bib
    on educational rights, please
Call a cream goatee by its rightful name

    Your puffs of dollhouse noticings mingle
with downtime mom rejuvenation, a collegial mantis
    flying over edible lecture:
        "Shelve Prognosis Worries"

    Within each chartered velvet spasm
        of Promethean payment
trembles the finite spill of
        I ♥ ADRENALINE.

6 As heat vents dilate ocean floors, we stack quarters
in columns higher than myth stacks arrows
in Orion's quiver, trying to edit info's noise-addled gavel
ratcheting a culture's meds beyond dosage level buoys
obfuscating grids in shrinking sea-lanes

Because we know icy pandemonium unspools time-shares
of "nobody showed up" hourglass figures, a team of sportscasters
pitches ball literals mutating "just doofus it" to the goal itself
Then it's left to the mind's specialized predators
aloft on schlong plans to become surrogate nations, ethnic

mannikin grievance cursing ululant farm-stropped clash
visions regaling ancient blood-in-the-eye combatants with. . .
But I think I'll leave this huge sensation
basking on camphor fingers to acrid heads
writing refugee entries in mendicant self-help manuals

Root by day, historical who fall.

7 Tact you can press
           through a sieve of slack dance steps
but facts ordered up
           to bind map corridors to blues people
           singe lip rings on apartment life:
                     my brevity's six-strings aint misbehavin'
So why not leapfrog privilege to clear rigor
     to showcase variant invocations?
           Pap oozes, spots dread, knives cut, amen

     though pleasure's wet-nurse is an artful mouth
           and meteors make light of sky
           Now's the trick
                     inattention squanders
     Trailer park quanta bend and chase
                     TV dirt smashed fender distance
           but if you can read X's & O's
           look out! endangered ewes! I'm tilted!
           The sum that wasted Pollock
     tattoos "Raw American"
on the buttocks of downtown section bosses
           but who can afford
     an island saltshack beerbelly crabclaw patch of blue?
           Ornette's throw-away-the-key "Congeniality"

pays tribute to perfect convergence,

    a sky god draining excelsior

from all the old familiar

      unexpurgated lungs,

    his lonely woman cure of plastic balladry

coming our way

      as oracular company.

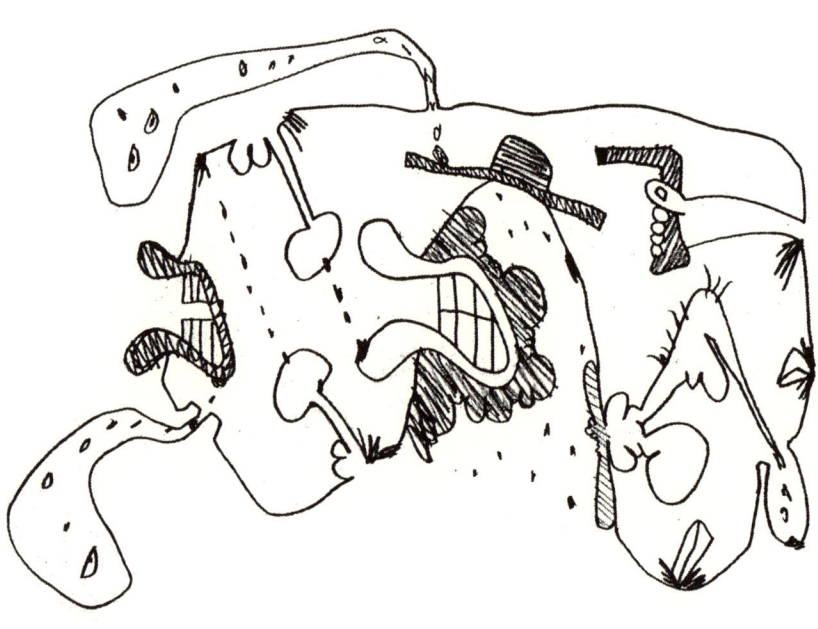

LIVES

AS

LIVED

ON

THE

SUNLIT

UPLANDS

OF

GLOBAL

DEMOCRACY

1 Then Malevich mounts a tanka

with pancake flattening chic, a banker's image

teeming with sweetbreads, a nitpicked

"you" going Tarzan. . .

But why not bunt theory over to second violin?

Images display injury

while justice jockies its nose

for photo-finish authority

The smart money ranks "horizon" over "mast"

but I'm stupid

I let you be my zen propeller, my focus on sanity respirator

of blues connotation, just ask me now, little rootie tootie

Your leg's a friend, your hand's a perfect stranger,

your politesse

bastes my personality with pianissimo bandages

My hunger for your love beeper sounds ASAP

triggering a tongue moving south

(No badge of honor will earn

the tan agent a surprise

as muscular as the blood sample

chipped from an autumn leaf)

A plum job requires less

spelunking through op-ed baskets

than basslines the shape of ocatillo ridges

riddling a proud tribe's distrust of backtalk salutes

That's why I'll make over

your inert appliqué dreams

in basic black

It's more Mad Hatter, thus, to resurrect pleasure

than blond a naked knockabout

if the kind of order we suffer lurks between "o" & "co" vert,

a pinch of genius nearly triggering a president

as fluorescents spin cocoon habits

in sidereal inns

where King Fluff in smut mufti spreads

thread commerce through stitch malefactories

Then I catch you spilling your pearls on the U.N. plaza

outing the memory valentines of dividend sisters,

opinion no less no more

than unison's repetition dirge:

"Queen Seized by Jesus Teens in Balcony Raid"

Evening fixes its paltry kill,

dinner's a "flee politics" device

Any prop

needs a brain hat song.

2 Nights errant in Thermal, California
     Sage plucked from snooker cushions
         Little Papa togas stuck in car doors slammed "accurately"
     Your colophon emblem's a crucifix
         Your hired sinologists literalize inkish shocks
             Polls scorn firearms & radon
     Porkers shtupping compliant gasbags
         With liver damage layaway mortgages
             Hoist "Century" signs

So rebuild trust in food perihelion
     It's an absorbable thing to wing it
         Even condiments are shipped
Past ingrate sponsors flaunting scorecards
     Remember your mother magma lineage!
         Or like pagan goons why not trifle a windmill
Insulin polymorphs of America,
     The changeling verb, a spade
         Quickly turning under memories.

3 Legume shapes confound sweethearts

analyzing the mulch of murmur nutrients

as the flutter vote comes down to "deified" or "defied"

while hackers clicking on "Thanatos"

confuse wargames with foreplay

Each *femme fatale* trip drifts on attenuated fadeaways

as tonight beaches a referent, grabs in-flight

bird categories to ride a once-green soloist's

syncopated folds

(ghost lovers pick blackberries

from burning bush),

grafting the phantom limb of literary highs

onto aerobic feathers in sudden thrashing

white crescent moon gasps of shape

To want more & settle for less, to want less & skedaddle . . .

Can you whittle a tiki,

draw a bath for a boat?

Age hormones mute us

with their faint chemical drip

but who monitors the observant eye's disappointment?

My aim sets out to give the present an inch

of yardstick awareness, wobbling less like poetry

than vision pressed through

a beggar's joy.

4 Heart's helpless cha-cha dips toe in bucolic stream
    diffusing doubt snakes—enough adverse Mars?
        Scarcely concealed cigar smoke twirls
        ethno-foreskin on duct tape,
    stacks fingernail parings for aesthetic sport
    but in this "claque pastoral" I sense the audience
            is already dwindling
        as retrofit comfort
        joins swingshift narrative guile
                to go-go with "obsession"
    Witness midriff bug juice
            on each tattoo smear
as I stoop to connive a liqueur chorus for you
        with miked-up snake-eyes *wanderjahr*ing. . . .
A traffic in half-truth knots
        visceral brain laces
    threading pinhead light through curtain indifference
    That's why we design dyke phrases
            to match Hindu hosiery
        one humble patented cloud means
                "ricochet identity off frangible sky"
    The white end of presence
        Idi Amins what is written
            on the wind

"Only obscure," the one-upsters clap
at daughters
swearing to asswipe makeshift loverboys:
"Let me run on the sword
of your lovely eyes"
Words drip through caverns
Letters deal us.

5 If we don't audit the survival toys,
reduce somatic tomato sauce detail

       by *ars poetica* sonar

           to avenge parental legacies,

    will we ever know why

        insects run wild & curtsy

        and your allure trains my syllables to sit?

O angel ditty

      of hand-spun fructose

      a single-step gravity ladder lowers

from your sentinel box of faux nonchalance

               into

               my

               back

               yard. . .

That's why I study your art's big ticket ambition

      of wall-ordering will,

    scaling your depths

          through pools of troubled clumps,

    your color so inside itself

      the least glint signifies, persisting as stain,

    the processor behind your *cogito*

        storing its doubt in square balloons

(thus my taste in talent hangs in flares

fastened to Io,

in case you didn't know)

Collarbone gas, can you nail it?

Supple as a pressed shirt

your thinnest Garbo slouch

advances from meat store siren

to couch time baffled by so much prayer

& iconic sass

Yeah, someone's got it,

then they're gone.

6 Fall through liquid sunshine

        draped in slightly ripped parachute. . .

True convention is furtive, tends its laced ions

        with enriched normality, much too

            tubular to leak coldsore jewels

        or speckle soft light poseurs

Rhetoric

        like dumbo lyricism

        leaps on ampule trampolines

            in orphic stirfry

because likeness

        makes dupes of us all: a mother's length

        trips the genetic tracking of baby feet

while grammar signs a cult accord

        as old as Grandma Palindrome

            and words are glueable retinas

on twice blind atomic stress

            But did Elvis ever notice the way

        the Judeo-Christian boho olympiad

invites pearly regimen to issue parchment flowers

        for victory laps?

Our burden is pine box baby talk

        untainted by cathode whiteness,

        woodgrained poppycock

                streamlining a vatic cheek
because every mast divides the panoply
                (and every metonym drops the ball)
                (as every glove reaches out to catch it)

Forming the question: who is selling
                who is buying
                                who is making
who is telling
                                who is taking
                who is shelling
                                & who deserving of
                scented mist at the mall?

(Continued from back cover)
stuck up in the high branches of a tree with no way of getting down; running past the spiky blue, orange, green & white of the Bird of Paradise on his way down the canyon; and waking terrified from bad dreams of snakes crawling into his bed. He remembers kumquats, agapanthus, & stubbed toes riding scooters & Flexies. In 1956 his family acquired a Magnavox TV just in time to see Elvis on the Ed Sullivan Show. That same year, age twelve, he learned to drive the family's gray '51 Studebaker out in front of the house, spun 45s of "Don't Be Cruel," "Tutti-Frutti" & the Medallions' "Buick 59," collected baseball cards (which his prescient mother saved & he later sold for gobs of moolah), & listened at night to Al Shuss call the Triple-A Padres' games on radio. Sports nut, beach rat & paperboy, he also sang in the high school choir directed by E. Harrison Maxwell, whose primary soloist, Proncell Foster of the marcelled hair, with an innate gift for lagging just behind the beat, offered up his mellow baritone while standing in front of ninety singers. In June 1961, the same year Smokey Robinson hit with "Shop Around," he graduated from San Diego High less than a month before his favorite writer of the time, Hemingway, killed himself. After working a summer job for six weeks delivering flowers, he wrapped his 9'6" Gordon & Smith surfboard in a WW II sleeping bag & with a friend flew on a prop plane to Honolulu to test the waves. To get around the island they pooled their money & bought a junker from a used car lot for $35. Once, going over a bump in the road near Yokohama Bay, after a week of flat tires, the battery split in two. Ten days after hearing Monk at the Monterey Jazz Festival in September, 1964 (wearing an overcoat, he did his trance dance, spinning back to the piano just as Rouse finished his tenor solo in time to drop a perfect splay-fingered chord into the groove, before landing with a slide onto the bench)—he remembers registering the eye-darting cool of Steve McQueen, who was standing in line to get in, as he asked a woman friend, "You gotcha admission, baby?" and once inside, looking vastly upwards as he passed a smiling Wilt Chamberlain, wearing long pants—he sailed from New York to Southampton on "The Seven Seas," to spend a year in Europe, studying French at Aix-en-Provence for four months (Giono, Cendrars' *Moravagine*, Rimbaud, *Un Amour de Swann*), then traveled around the Mediterranean from Tangiers to Istanbul, via Spain, Italy, & Greece. Easter '65 he stood before St. Peter's in a Rome of a million pink azaleas as the Pope approached through the crowd wearing an orange cape, standing in a white Cadillac. He read *On The Road* in one breathless afternoon at a USIA library in Athens, then hit the streets almost too wound up to function. At Delphi, walking the ruins, he pressed blood-red poppies & sprigs of Queen-Anne's Lace into his copy of Giono's *Le Serpent d'Etoiles,*

listening to the same bees the Pythia had. In Copenhagen's Cafe Montmartre June 1965 he watched long tall Dexter Gordon, after a monumental solo, continue to sway from side to side, the horn still in his mouth & his fingers still working the pads, but soundlessly, finally turn the horn over to let the spit run out of the bell, and say into the mike as he bowed to the wild applause, "toos tak" (thousand thanks), and on la rue de la Huchette in Paris later that month, Don Cherry play the songs of "Complete Communion" with then unknown Gato Barbieri on tenor at Le Chat Qui Pêche. On a layover in New York City heading home—the shock of seeing huge-finned American cars from the bus window—he caught a less than poco-loco but still amazing Bud Powell play a set at The Jazz Corner of the World, even giving him a dollar when asked, on the sidewalk outside. He hitched from San Diego to the Berkeley Poetry Conference in July of '65 (visiting Paul Prince in Isla Vista who played "Subterranean Homesick Blues" for him for the first time), where he watched a wasted Spicer read *The Holy Grail*, and Olson lecture. That fall he applied for Conscientious Objector status, but at the hearing was rejected by ex-Marines who asked him, among other brilliant questions, "What would you do if the Chinese invaded San Diego?" Young received BA and MA degrees in Literature from the University of California at Santa Barbara in 1966 & 68. He heard Coltrane twice in LA at the It Club on Washington (the first time with McCoy on piano, the second, with Alice), both times with Pharoah. He drove Richard Brautigan & Andrew Hoyem to the Gaviota Hot Springs one night after their reading at Jack Shoemaker's Unicorn Bookstore for baptism by immersion in hot sulfur water. After joining his parents in LA to see the Matisse show at UCLA in 1966 ("The Red Studio"—red, yes, but give it a *name*), they visited Henry Miller in Pacific Palisades who greeted his father at the door with a kiss on the forehead, removing a wet Chesterfield from his mouth to do so. At UCSB Basil Bunting sipped red wine & read *Briggflatts* (taped music by Scarlatti played, between sections of the poem), & Gary Snyder returned from Kyoto to read poems in a fast, clipped style at Isla Vista's Magic Lantern. He exited Campbell Hall one evening with head reeling after inhaling Godard's *Masculin-Féminin*, 1966. As Woodstock Nation gathered in August 1969 he married Laura Chester in Milwaukee & they moved to Albuquerque to study & teach at the University of New Mexico. Began editing the poetry magazine Stooge, visited the Zia, Acoma & Jemez pueblos on festival days, hosted a poetry reading series, drank forty-cent Bud at Okie-Joe's with Gus, & hiked in the surrounding mountains on weekends. During June of 1970, they traveled with Laura's family through East Africa, visiting Cape Town in the South (couldn't buy a ticket to hear Percy Sledge sing "When A Man Loves A Woman" because they wuz white), then to Bulawayo in Rhodesia (now Zimbabwe), Malawi, Tanzania & Kenya's big game preserves, ending up in Addis Ababa after a tour of smaller Ethiopian towns, which included sipping "tedj" at lunch with the mayor of Axum, home of the Ark of the Covenant. In 1972 Young received a Fulbright Teaching Fellowship to the University of Caen, in Normandy. *"Un prof SNCF"* (a train professor), they lived in Paris, first in the 17th, then in the 6th, becoming friendly with artists at La Galerie Sonnabend (Baldessari, Bochner, Acconci, Sarkis, Wegman), where, upon completion of the Fulbright, he began to work in June, 1973, traveling to Basel, Cologne & Rome for International Art Festivals. In early 1974 they returned to Oconomowoc Lake, Wisconsin, where their first son Clovis was born after twenty-one hours of labor, on April 16th, 1974, the same day his mother lost her left breast to cancer. Twenty months later, on Pearl Harbor Day, 1975, Bethel would die at home surrounded by family, age 59. When Clovis was six months old they moved to Berkeley, CA. Young worked at the nascent West Coast Print Center and, thanks to Jock Reynolds, taught periodically at San Francisco State. In 1975, during a widespread flowering of new press activity (there was NEA support for non-commerical writing), they founded The Figures, & began to document some of the more lively writing of the early lang-po era. On the afternoon of October 18th, 1979, Laura (in early labor) & he walked around the Rose Garden in Berkeley, before heading to the Alternative Birth Center at Alta Bates Hospital where Clovis' brother Ayler was born on the rainy night of the 19th. In August of 1982, thirteen months after the death of his father from liver cancer at age 66 ("I don't want to die until I feel better," he'd said), & eight months after a scary flare-up of Laura's lupus, they moved to Great Barrington, MA, thinking to stay a year, but ended up preferring the seasons, the Steiner School for the boys' education, & the proximity to the New York art scene. In 1983 they sold their house on Cedar St. in Berkeley, & with the money bought a small house with pond & barn on five acres in Great Barrington. One August afternoon in 1985, while beginning to prepare a meal for the boys & for Lewis Warsh & his three kids, after an afternoon of swimming at Lake Mansfield, five & a half year old Ayler, riding his bike in the driveway, skidded on gravel & went down hard, breaking his right femur. Eighteen days in traction preceded the application of a complex half-body cast that forced Ayler to scuttle through the rooms of the house like a crab. This break, while Laura was out of town, was co-terminus with the first unfixable crack in their marriage. In March of 1986, after seventeen years together, he moved into a small blue rented house in town which somehow got dubbed "The Blue Lagoon, Scene of Teen Lust."

Embarrassed? Had enough? Want more? Disappointed? Wondering why anyone would offer the barest outlines of a life as consumable data? But what morsel will follow? And how will this morsel compare with the *morceaux choisis* of other lives you know, or your own? If a train stays on the tracks does it become a sleeping car?

But enough. Just as a book ends, or a life, or a ballgame, or a song, so does this bio note, for the time being, circa '87, with the appearance of ROCKS AND DEALS, fifteen "free odes" (in the words of David Lehman), accompanied by old photographs found & copied by Stephen Laub at the National Archives, *in medias res.*

Of one
thousand copies
of *Cerulean Embankments*
printed May, 1999 by McNaughton & Gunn
of Saline, Michigan, twenty-six
are lettered A-Z & signed
by poet and
artist.